Karen Romano Young

Arctic Investigations

Exploring the Frozen Ocean

RSVP

RAINTREE
STECK-VAUGHN
P U B L I S H E R S
A Steck-Vaughn Company

Austin, Texas

www.steck-vaughn.com

For Nancy and John,
with love and warm wishes

Steck-Vaughn Company

First published 2000 by Raintree Steck-Vaughn Publishers,
an imprint of Steck-Vaughn Company.

Copyright © 2000 Turnstone Publishing Group, Inc.
Copyright © 2000, text, by Karen Romano Young

Library of Congress Cataloging-in-Publication Data

Young, Karen Romano.
    Arctic investigations: exploring the frozen ocean / Karen Romano Young
p. cm. —(Turnstone ocean pilot)
Includes bibliographical references.
    Summary: Examines past and present scientific study of the Arctic, describing what life is
like for scientists staying there and explaining how and what they study.
ISBN 0-7398-1232-7 (hardcover)   ISBN 0-7398-1233-5 (softcover)
    1. Arctic regions Juvenile literature. 2. Scientific expeditions—Arctic regions Juvenile
literature. [1. Scientific expeditions—Arctic regions. 2. Arctic regions.]
I. Title. II. Series.
0614.Y68 1999   99-23541
919.804—dc21   CIP

For information about this and other Turnstone reference books and educational
materials, visit Turnstone Publishing Group on the World Wide Web at
http://www.turnstonepub.com.

Photo credits listed on page 48 constitute part of this copyright page.

Printed and bound in the United States of America.

1 2 3 4 5 6 7 8 9 0 LB 04 03 02 01 00 99

Contents

1 Out on the Ice

At the top of the world is a vast sheet of ice. In the winter ice covers an area about the size of the United States. In the summer the ice is half as large.

In every direction there's nothing but shades of white and blue. The air is clear and clean. The only sound is the singing, sighing, and moaning of moving ice. And it's cold, very cold. Being at the top of the world is not like being anywhere else on Earth.

The Arctic is the name given to the whole area inside the Arctic Circle, an area around the top of the world that includes the northern parts of Greenland, Canada, Alaska, Russia, Finland, Sweden, and Norway. In the center of the Arctic Circle is the Arctic Ocean, the only ocean in the world that's covered with ice.

Go North!

How do scientists get to the Arctic? It's simple. They go north. That's how everybody gets to the North Pole, isn't it? But there are lots of different ways to travel.

Scientists heading for the Arctic can drive, fly, or travel by sea to towns on the coasts of Sweden, Russia, Canada, or Greenland. Then, they board boats for the open ocean or fly by plane or helicopter out to an ice camp. Or they stay aboard ship and conduct experiments in the water.

BERING SEA

Alaska

ARCTIC CIRCLE

ARCTIC OCEAN

Permanent Ice

Canada

NORTH POLE

Russia

Greenland

Winter Ice

Norway

Finland

ATLANTIC OCEAN

Iceland

Sweden

It's ice that makes the Arctic a home for tiny plants and animals and a feeding ground for fishes, birds, bears, and whales. It makes freshwater to add into the rest of the world's oceans. Ice is the first and last thing that most people who come to the Arctic see. It covers the Arctic Ocean in sheets that average about 2 meters (about 6 ½ feet) thick. It's the ice that makes scientists eager to visit the Arctic—and nervous about what it will be like to work there.

Getting to the Arctic isn't easy. It's so cold there that engine parts sometimes freeze. But scientists depend on the supplies the airplanes deliver. Before an airplane can take off again, its engine must be warmed by gas-powered heaters. Heat thins the engine's oil, which becomes like gelatin in the cold.

Scientists who have traveled to the Arctic ice all remember one moment—when the plane that brought them took off and left them behind. Jeff Lord describes that moment with one word: "silence." Jeff organizes Arctic trips for scientists at the Woods Hole Oceanographic Institution (or WHOI, pronounced "HOO-ee") in Woods Hole,

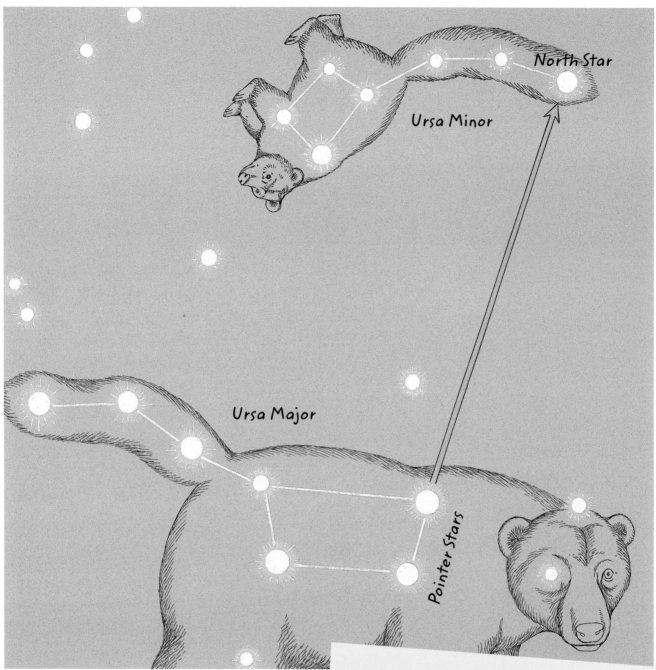

North Star

Ursa Minor

Ursa Major

Pointer Stars

Massachusetts. Even though Jeff has been to the Arctic many times, he still gets a chill when the plane leaves—and it's not just from the cold air.

That moment is chilling, but it's thrilling, too. For every scientist who arrives in the Arctic, there's a feeling of, "I'm here at last!" It takes a long time to

The name Arctic comes from the Greek word *Arktos*, which means bear. Above the Arctic is a constellation, or group of stars, called Ursa Major. (*Ursa* is Latin for bear.) The constellation is also called the Big Dipper. Two stars in this constellation act as pointers to a very bright star called the North Star, which is part of the Little Dipper, or Ursa Minor, constellation. The North Star is almost directly over the North Pole. It has been used by sailors for thousands of years to mark which direction is north.

This is pack ice, ice that forms over the ocean. Though in the Arctic the pack ice is thick, it can still sometimes crack, and that can be dangerous.

get to the Arctic. Plenty of planning, dreaming, and hard work lead up to that moment when the plane takes off, leaving the scientists hundreds of miles from any other humans. Scary? Yes. Dangerous? Definitely. There are no cities or towns at the ice camp, no hotel or friend's house to stay in, not even an airport. (In fact, a visitor's first job in the Arctic may be to build a landing strip for airplanes.)

Step Carefully

The need for careful preparation was something Keith Von Der Heydt, a WHOI engineer, felt as soon as he was on the Arctic ice. "I realized right then that it was important to do the right thing," says Keith, who has made nine trips to the Arctic.

Most of Keith's work takes place in ice camps, little towns of tents pitched on the ice. Ice camps allow work to be done in places that can't be

Keith Von Der Heydt (standing on the left, with Eddie Scheer, another WHOI engineer) designs machines and instruments that help scientists do their work.

reached from a ship. Things happen in an ice camp that scientists don't have to worry about anywhere else. They're not on a ship, a beach, a mountaintop, or a piece of land. They're on floating ice, which melts and freezes and cracks and crunches, forming big ice mounds and ridges. "If you make a series of mistakes on the ice, you may not recover," Keith warns.

Staying on the ice is a big problem for scientists at an ice camp. "You don't absolutely know the camp won't break up," says Keith. A crack can open in the ice under a scientist's bed, or a submarine could surface, breaking through the ice. Once, a tent slid into a crack when the ice beneath the camp split. An entire camp can float away, or an airplane

It's tricky camping on ice. A big crack in the ice has opened up right beside the camp. If the crack gets large enough, part of the ice pack can drift off as an ice floe, a piece of pack ice that breaks and floats away.

Snowmobiles are one way to get around in the Arctic. Here, Jeff Lord (left) and Kim Pelle are bringing in a generator to provide electricity for an ice camp.

can fall through the ice—things that have happened at least once. Scientists must be alert in the Arctic because they never know what might happen.

Because surprises can always be expected in the Arctic, it's important that conditions at an ice camp be carefully considered. This includes planning for emergencies. That's why having someone like Jeff Lord along is so important. "My job is to figure out anything you need to live for six to ten weeks on the ice," Jeff says. He looks at the scientists' plans, then figures out the equipment, clothes, food, and transportation they'll need. Jeff is also an emergency medical technician, prepared to do all kinds of first aid, such as binding a sprained ankle or even saving someone's life. If there's an emergency in the ice camp, Jeff is the one to call.

Scientists in the Arctic don't work a day without thinking about other people who have explored there, studied there, and—sometimes—died there. In the Arctic you have to work quickly, step carefully, and watch out for danger.

Bear Scare!

In an ice camp, the biggest threat can have four feet, white fur, and a big, black nose. Polar bears are naturally curious, and they get hungry when they smell food. Bears can pop up at any time, sniffing out the camp's dinner or just finding out what's going on.

Keith came across this huge bear track near an ice camp. Polar bears can be as tall as 3 meters (about 10 feet) and weigh up to 500 kilograms (about 1,100 pounds). Caution is always needed when large predators like polar bears are in the area.

2 Ice Pioneers

Scientists working in Antarctica, the continent opposite the Arctic that includes the South Pole, wake up each day in the same place on the globe. That's because Antarctica is ice-covered land, not ice-covered water. But scientists doing research on the Arctic ice are always on the move. Arctic ice drifts, moved by currents, or flows of water, that push water and ice. Arctic ice moves slowly over the North Pole over the course of a year. Riding along, ice-camp scientists are at a different place every day.

People didn't always know the ice moved. They thought the Arctic ice was frozen and still. Then, in the late 1800s, a Norwegian scientist named Fridtjof Nansen heard about Siberian driftwood that had washed up on the shores of Greenland, halfway around the world. How could that happen? Nansen thought that there must be a current and that the current had pushed the ice, along with the driftwood frozen into it, from one side of the North Pole to the other.

Before 1893 most people who traveled to the Arctic wanted to plant their country's flag there. They were explorers who hoped to be the first to reach the North Pole or the first to find a path through the Arctic ice. Nansen wanted to get to the Pole, too, but along the way he wanted to find out more about the Arctic itself.

Nansen had always been interested in the Arctic, but he knew that a trip there meant hundreds of miles of dangerous walking, dogsledding, and skiing. Many people had died trying to reach the North Pole, killed by cold, hunger, injury, or even polar bears. If only there was a ship that could travel through the Arctic ice. But how could a ship move through ice?

The driftwood made Nansen think. If the current could move driftwood, maybe it could move a ship, too. He decided to try to prove his belief that there was an Arctic current and, at the same time, to try to win the race to the North Pole.

Nansen had a ship, which he named *Fram*, specially designed for a trip through the ice. Braced, or supported, with extra timbers, the ship was strong enough to withstand the crush of Arctic ice. Here are Nansen and *Fram* on their journey through the ice.

Locked in the Arctic Ice

In 1893 Nansen set out from Norway in a ship called *Fram*. He sailed the ship to Siberia, then moved it up to the edge of the Arctic ice. As the ocean froze that fall, *Fram* was frozen into the ice. For the next three years, Nansen and his crew drifted, stuck in the ice. They cooked, ate, slept, and stayed warm in the ship. Outside they sunbathed on bright days, did a little hunting, and fought off polar bear attacks.

Every two to four hours for three years, *Fram* crew members recorded water temperatures. They used bottles that worked in a way Nansen invented to collect water samples. The bottle was such a good

Arctic Exploration

Early Settlers

1000
Thule Eskimos crossed the Bering Strait from northeast Asia.

Search for a Passage Through the Arctic

1476
Explorers from England and Holland began to look for passages through the Arctic, without success.

| 4000 B.C. | 0 | 1000 A.D. | 1400 | 1500 |

4000
The Inuit arrived in North America, traveling from Greenland.

1100
Norse explorers settled in parts of Greenland in the Arctic.

invention that it was still being used almost one hundred years later.

Every day Nansen also wrote in *Fram*'s logbook. Every ship captain keeps a logbook, or log, but Nansen's log had many extra details. He recorded not only where the ship was, what the crew did, and water and air temperatures, but also what the crew ate and who cooked. He even included poems and songs the crew wrote. There were two reasons for this. First, he wanted to be able to show his family and friends what he'd been doing. While *Fram* was in the ice, there was no way to send a message to anyone. Nansen's log had a second purpose, too. It was also a record of everything he learned or thought about the Arctic.

1700s
No Arctic passage had yet been found, but as fur trading and whaling grew, more of the Arctic was mapped. Explorers, such as William Baffin and Vitrus Bering, sailed the Arctic. The waters and islands they found were given their names.

1878–1879
Success! Adolf Erik Nordenskjold was the first to navigate the Northeast Passage.

1700   **1800**

1845–1847
Sir John Franklin of England set out to find a way through the Arctic called the Northwest Passage. The expedition was never heard from again. It was years before any trace of the lost ships and people was found. Franklin's body was never recovered.

Race to the North Pole

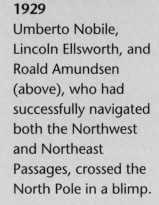

1907–1909
Several explorers claimed to have reached the Pole during this period of time, including the team of Admiral Robert E. Peary (left) and Matthew Henson (right) of the United States. To this day there is no proven claim of reaching the North Pole on foot using dogsleds.

1926
Admiral Richard E. Byrd claimed to have flown over the North Pole. His pilot, Floyd Bennett, denied getting there.

1929
Umberto Nobile, Lincoln Ellsworth, and Roald Amundsen (above), who had successfully navigated both the Northwest and Northeast Passages, crossed the North Pole in a blimp.

1900

1930

1893–1896
Fridtjof Nansen crossed the Arctic Ocean on a ship frozen into the ice. He also attempted to reach the North Pole.

Nansen designed sampling bottles that would flip end over end at the desired depth and then fill up with water. "Nansen bottles" were still being used one hundred years later to take samples of water at different depths.

1930s
Careful, orderly research of the Arctic waters near Russia was begun by the former Soviet Union. Expeditions started each year as soon as the ice allowed and continued until the ice closed in again in the fall.

Science in the Arctic

Modern Exploration

1958
The nuclear submarine *Nautilus* made the first under-ice crossing of the North Pole.

1959
A U.S. stamp was issued to commemorate fifty years of Arctic exploration.

1977
Arktika, a Russian nuclear-powered icebreaker, became the first surface ship to reach the North Pole.

1960

1990

1937
The former Soviet Union set up the first floating scientific station. The station was established near the North Pole and drifted south for nine months until it reached the Greenland Sea.

1957–1958
Both the former Soviet Union and the United States began Arctic research programs.

1969–1975
The United States and Canada joined in a research program to study how Arctic sea ice responds to the environment.

1997–1998
Des Grosseilliers (above), a Canadian icebreaker, was frozen into the Arctic ice during a global research mission called SHEBA.

17

Early Arctic exploration was dangerous. Many ships became trapped in the ice and were crushed. Nansen's *Fram*, shown above frozen in and drifting with the Arctic ice, was strong enough to survive the pressure of the ice.

Science in the Arctic

In August 1896 *Fram* broke free of the mushy summer ice and sailed south. The ship landed close to Spitsbergen, an island north of Norway. Nansen was proved right—*Fram* had started drifting in the ice in Siberia and had ended up on the other side of the Pole.

Using his log, Nansen was able to prove that at the top of the world there is a deep, moving ocean, not a shallow lake or sea. Nansen also realized that the Arctic Ocean has a special role in the world. It sends cold freshwater flowing into the Atlantic Ocean. He didn't know why it happens or how it happens, just that it does. Maybe the most important thing Nansen accomplished was to show scientists that the Arctic region was a place that needed to be studied.

When Nansen returned, he was called the first polar oceanographer. He was the first person to measure many things that oceanographers check closely today. Thanks to Nansen, oceanographers can compare what happens in the Arctic now with things that happened a century ago.

Today scientists go to the Arctic to learn more about the region's ecosystem—how living things there exist with the ice, water, winds, and currents. They also want to know how what happens in the Arctic affects the rest of the world. To find out, scientists study not only the Arctic's plants and animals, but the Arctic itself.

A hundred years after *Fram*'s voyage, a Canadian ship was frozen into the ice. Like Nansen, the scientists aboard *Des Grosseilliers* were there to investigate mysteries at the top of the world.

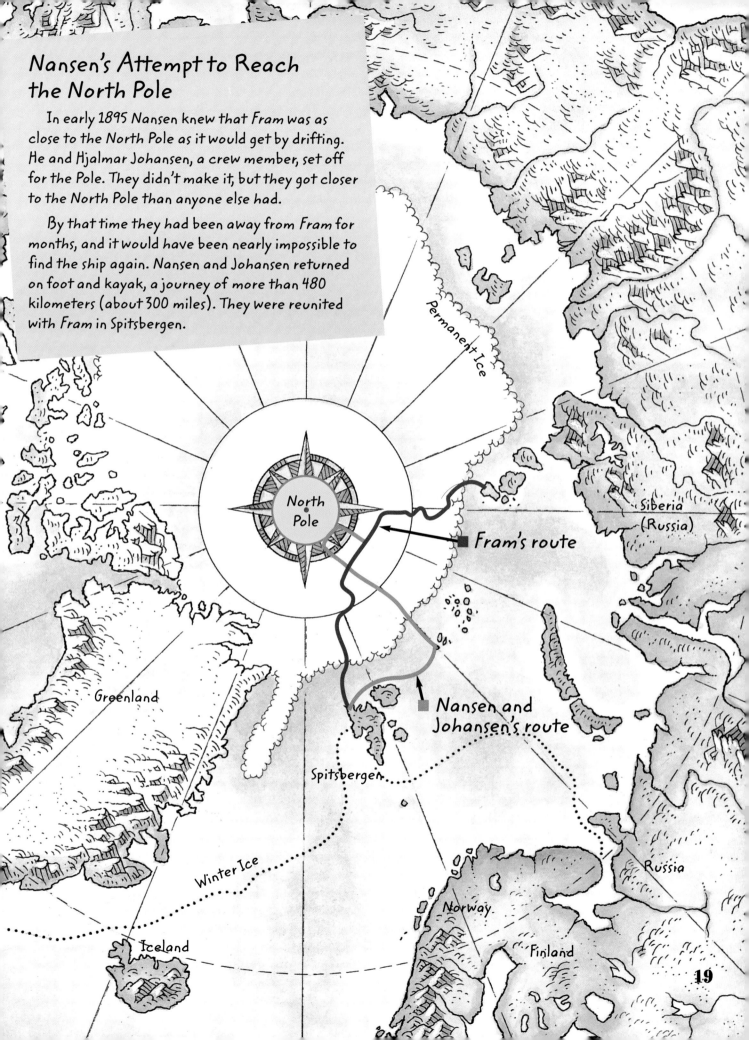

Nansen's Attempt to Reach the North Pole

In early 1895 Nansen knew that *Fram* was as close to the North Pole as it would get by drifting. He and Hjalmar Johansen, a crew member, set off for the Pole. They didn't make it, but they got closer to the North Pole than anyone else had.

By that time they had been away from *Fram* for months, and it would have been nearly impossible to find the ship again. Nansen and Johansen returned on foot and kayak, a journey of more than 480 kilometers (about 300 miles). They were reunited with *Fram* in Spitsbergen.

Permanent Ice

Siberia (Russia)

North Pole

■ Fram's route

■ Nansen and Johansen's route

Greenland

Spitsbergen

Winter Ice

Russia

Norway

Iceland

Finland

3 Frozen In

Des Grosseilliers was a floating laboratory and home for more than fifty scientists from all over the world for 13 months during a global research mission. The mission was part of an eight-year study of the Arctic sponsored by organizations in the United States, Canada, and Japan.

In September 1997 more than one hundred years after *Fram*'s journey, *Des Grosseilliers*, a Canadian icebreaker, steamed into the Arctic and waited to be frozen into the ice. By autumn the Arctic is not a place where most people want to be. The earth has tilted until the North Pole is almost completely turned away from the sun. Each new day gets colder. The sun rises later and later and sets earlier and earlier until the sky is dark nearly all day. As it gets colder, water at the edge of the ice freezes, and the ice pack gets larger. It's a time when it's easy to believe that the Arctic is a dead place where nothing grows.

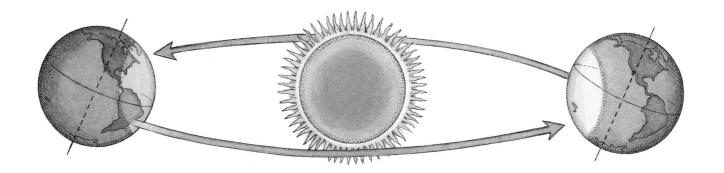

But the land and ice in the Arctic Circle are full of living things—from whales, walruses, and polar bears to smaller seals, foxes, rabbits, fishes, and insects. There are also plants and animals so tiny that they can be seen only through a microscope. How these plants and animals live in the cold of the Arctic is something that interests scientists.

The scientists on *Des Grosseilliers* were part of a project called Surface Heat Budget of the Arctic (or SHEBA). The purpose of SHEBA was to find out if the Arctic was becoming warmer. If it was, the scientists wanted to know why it was happening and what warming would mean for the plants and animals in the Arctic. They would also try to determine if changes in the Arctic ecosystem would mean changes all over the planet.

For years some scientists have predicted that pollution will make Earth warmer. Carbon dioxide from cars and other machines and chemicals released by air conditioners, refrigerators, and some spray cans could build up in the atmosphere. These gases, called greenhouse gases, could trap sunlight and warm the globe. Some of these gases could also

Why is the Arctic cold? Mainly it's because the sun never shines directly on the North Pole, due to the tilt of the earth. The heat that does reach the Arctic pack ice is reflected back by the white of the ice. So even in summer the ice doesn't melt much. In the winter the ice stops warmth from rising from the ocean into the air. This keeps the air cool and helps even more water to freeze. The extreme cold near the surface of the ice causes the ice to thicken. Water temperatures near the surface can be as low as -2°C (28.4°F).

Here, SHEBA scientist Igor Melnikov of the Shjirshov Oceanography Institute in Moscow is getting ready to dive under the Arctic ice. Once there, he'll take photographs of organisms that live on the underside of the sea ice. This will help scientists study seasonal changes in creatures that live in the ice.

make part of the atmosphere called the ozone layer become thinner.

The ozone layer is made of ozone, a gas that absorbs certain rays from the sun. If the ozone layer became thinner, more of these rays would reach Earth. These rays could gradually warm Earth and cause harmful changes in living things.

One serious result of warmer temperatures could be changes in global weather patterns. Farming areas could become too dry to grow crops. Warming could also cause the Greenland and Antarctic ice caps, thick masses of ice and snow that cover an

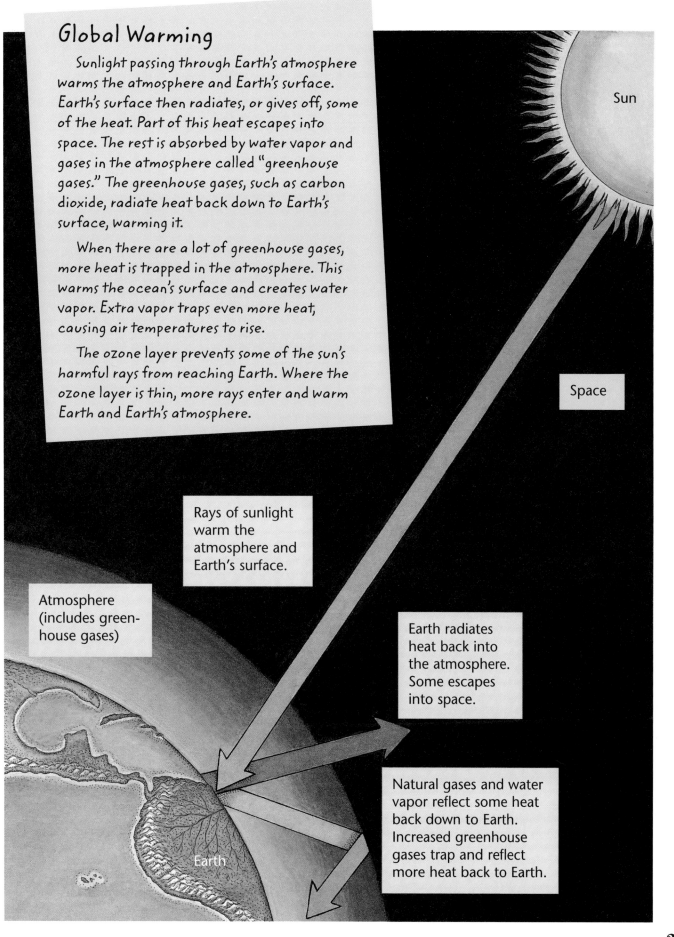

Global Warming

Sunlight passing through Earth's atmosphere warms the atmosphere and Earth's surface. Earth's surface then radiates, or gives off, some of the heat. Part of this heat escapes into space. The rest is absorbed by water vapor and gases in the atmosphere called "greenhouse gases." The greenhouse gases, such as carbon dioxide, radiate heat back down to Earth's surface, warming it.

When there are a lot of greenhouse gases, more heat is trapped in the atmosphere. This warms the ocean's surface and creates water vapor. Extra vapor traps even more heat, causing air temperatures to rise.

The ozone layer prevents some of the sun's harmful rays from reaching Earth. Where the ozone layer is thin, more rays enter and warm Earth and Earth's atmosphere.

Sun

Space

Rays of sunlight warm the atmosphere and Earth's surface.

Atmosphere (includes greenhouse gases)

Earth radiates heat back into the atmosphere. Some escapes into space.

Natural gases and water vapor reflect some heat back down to Earth. Increased greenhouse gases trap and reflect more heat back to Earth.

Earth

Scientists Donald Perovich and Edgar Andreas of the United States are starting to drill down into the ice. They are standing on a little more than 1/2 meter (about 2 feet) of ice and 3,300 meters (about 11,000 feet) of water. SHEBA scientists used small hand-guided drills like this one to determine the thickness of the ice.

area of land year-round, to melt. This melting would make sea levels rise. People and homes on coasts everywhere could be threatened by rising water. Because the ozone layer is already naturally thinner at the Poles, it would become warmer there first. So the Arctic is a place where scientists can look for early clues.

Arctic Measurements

But is global warming really happening? A good way to answer that question would be to go to the Arctic and make measurements. One of the SHEBA scientists' first jobs was to measure the thickness of the Arctic ice. Thinner ice than usual would mean more ice had melted. That would mean the Arctic was warmer. Inside their big, red ship, the SHEBA scientists pulled on heavy parkas and boots and went out to measure the ice.

They were shocked by what they found. The ice was only 1.2 meters (about 4 feet) thick in places. The ice should have been at least 1.8 meters (about 6 feet) thick. They also found that the ice was freezing later in the year and melting earlier than in past years. Some of the measurements SHEBA scientists used to compare past and present conditions were Nansen's, taken during *Fram*'s voyage. SHEBA scientists also used satellite data. Since the 1970s, satellites orbiting Earth have taken pictures and made measurements of the Arctic ice. All the data SHEBA scientists used pointed to a warmer Arctic.

While the SHEBA scientists agreed that the Arctic was warmer, they couldn't agree on why. Is the world just in a naturally warm time? After all, the world has been warming slightly and cooling slightly, over and over, for millions of years.

Is pollution making the temperature warmer, or is it simply making a naturally warm time warmer? What would a warmer Arctic mean for the plants and animals that live there—and for the rest of the world? SHEBA scientists had a lot of questions. They hoped to find some answers during their long stay in the Arctic ice.

Even with airplane deliveries, modern communications, and the company of other scientists and crew, the SHEBA mission meant long hours in some very lonely places. But everyone on the cruise was excited about the chance to work in one of Earth's most challenging places.

Down into the Ice

The SHEBA researchers took ice cores using small hand drills. This helped them determine the thickness of the Arctic sea ice and if the Arctic is becoming warmer. On an ice cap, the ice is much thicker, so researchers use a large motorized drill to take cores.

Ice in an ice cap stacks up, year by year and century by century. The ice deep beneath the surface can be thousands of years older than the ice above it. By studying ice of different ages, scientists can learn about conditions at the time the ice formed. Cores from an ice cap in Greenland can provide information about conditions in the Arctic long ago.

1

First, scientists choose a place to drill. Next, they set up camp nearby, but far enough away that the ice is undisturbed. Then, equipment for drilling, studying, and storing the cores is brought to the site. The biggest piece of equipment is the support tower for the pipe that will go down into the ice. The white dome that houses the drill is about 15 meters (about 50 feet) tall. The tower extends from the top of the dome about 30 meters (about 100 feet).

2

Once everything is set up, it's time to drill. Here, the winch motor slowly lowers a drill attached to a cable inside the tower. The drill is connected to a device called a core barrel, a hollow pipe that fills with ice as the drill spins. The drill and core barrel are lowered into a hole in the ice. The hole has been filled with a liquid that keeps the hole from closing up over the drill. Computers are then used to operate the drill as it moves deeper and deeper into the ice.

3

Once the drilling has finished, the core is pulled back up to the surface and removed from the core barrel. In the drill dome, a powerful saw is used to cut the core into smaller lengths that are easier to handle.

4

The core is then cut into even smaller pieces. Part of the core is put into a large freezer to store for future experiments.

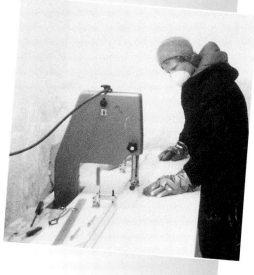

5

The ice core samples are studied in many different ways. Scientists may check the chemical content of the ice or look at soil particles trapped in the ice. The chemicals and soil in an ice core can tell scientists about conditions in the Arctic a long time ago.

6

This is an ice core section from the bottom of a glacier, a huge mass of moving snow and ice. It shows a combination of ice and soil particles.

4 Reading the Signs

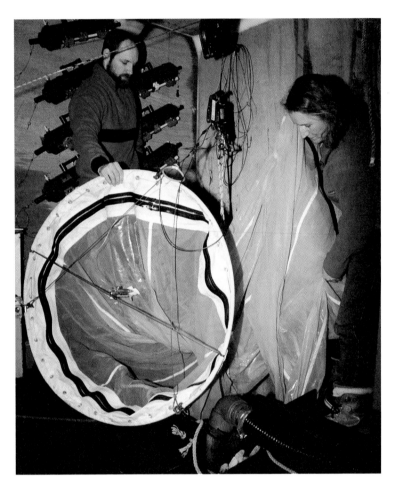

This net was pulled through the water by a cable attached to a big winch. The net scooped up copepods, the animals studied by Robert Campbell of University of Rhode Island (left) and Carin Ashjian.

Some scientists study the ice in the Arctic, while others study the plants and animals that live in that ice. Some of the animals are huge, such as whales and polar bears, and some are very, very small. Carin Ashjian, a WHOI biologist, studies a tiny cousin of lobsters and crabs called a copepod. During SHEBA she worked in a tractor-trailer container next to *Des Grosseilliers.* Carin used a net to collect copepods, then stored them in a tank of water so cold that it hurt her fingers to dip into it.

Later, back home in Woods Hole, Carin determines how far under the water's surface and at what temperature each kind of copepod lives. She also checks each copepod to see how much fat it has. The amount of fat is a clue to how much food the animals are finding to eat, and that can help Carin know how well the copepods are surviving.

Learning how these animals, the water temperature, and the water depth are connected helps SHEBA scientists learn more about the copepod's place in the Arctic ecosystem. That's

important to know because many different animals eat copepods. If the copepods are well fed, it may mean they are thriving and plentiful. That means the animals that eat copepods may have enough to eat, too. But if water temperatures change and cause copepods to move to a different depth, it may mean that hungry animals won't be able to find them. Studying copepods can help scientists understand how warming temperatures may change life in the Arctic.

Knowing that a kind of animal is found in an ecosystem tells you that the animal is able to survive

"In the winter you look out and see nothing but snow. But there's a whole life underneath the ice that you can't see," says biologist Carin Ashjian. Here, Carin is checking on the copepods she's collected in her net. Copepods are tiny. The copepod below is really only this long: H.

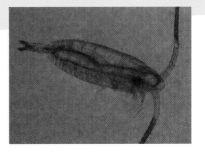

Oceanographer Cynthia Tynan explains, "We still have many gaps in our information about the Arctic. If global warming continues, the food supply for whales may change, and whales may go thousands of miles to get to new feeding places."

in that place. And if there are a lot of one kind of animal in that ecosystem, it shows that it is a good place for that kind of animal to live. When scientists pay attention to which animals live in the Arctic, how many kinds of animals there are, where they live, and how they're doing, scientists are reading signs. Signs point to what the Arctic ecosystem is like, how healthy it is, and how it's changing.

Whale Watching

SHEBA scientists aren't the only scientists who study the problem of warming temperatures—and it's not always tiny animals that they study. In the Arctic, Cynthia Tynan stands on the bridge of a ship looking through the lenses of a pair of giant binoculars. Cynthia is a biological oceanographer at the University of Washington in Seattle, Washington. She studies where whales are found by surveying whales over vast ocean areas.

When she surveys whales, Cynthia records the number and kinds of whales that she sees. She compares this data with information about the water, such as temperature and saltiness, and the amount of prey, or food for the whales, in the water. She then compares where whales are found now with where they have been found in the past to see if patterns have changed.

Animals such as whales rely on Arctic waters for prey. If the place where their prey lives changes or if there are fewer prey than usual in a place, the whales will move to another place. As the Arctic warms and there is less ice, the numbers and kinds of plants and animals in the Arctic ecosystem may change, too. This may cause changes in the movements or migration patterns of animals like whales as they hunt for food in new areas. Research is under way to learn what these changes might mean for Arctic animals. Ecosystems are very complex, and one change can cause many more.

For most animals, finding enough food is a very important part of their lives. Every kind of animal, including humans, is part of a food chain. A food chain starts with a plant or an animal. Sometimes the start, or bottom, of a food chain is dead plant or animal matter. Something eats it. Then, something larger eats that.

Whales, polar bears, and walruses are examples of animals at the top of food chains. No one eats them—except, maybe, people. They are also examples of indicator species. An indicator species is a living thing that can provide information about the place it lives. The health and well-being of an indicator species—as well as how many of them there are in an area—can be clues to how well animals farther up or down a food chain are doing. For example, if the number of whales in a certain area decreases, it may mean that there is less prey for them to eat. Indicator species like whales can serve as an early warning that the health of a community or an ecosystem is being threatened.

Humpback whales feed on schools of fishes and tiny animals called krill. If the whales move to a different part of the ocean, it may mean that the fishes and krill have moved, too. What Cynthia learns about whales also provides her with information about their prey.

An Arctic Food Chain

Each animal in a food chain depends on a plant or another animal for food. If something happens to one plant or animal in the chain, all the animals in the chain can be affected.

Polar Bears

Polar bears are the largest land meat-eater, and their main food source is a kind of seal called a ringed seal. If ringed seals are available, a polar bear will eat a seal every few days. If ringed seals are hard to find, polar bears can go weeks between meals.

eaten by

Ringed Seals

These short, plump seals are the most numerous mammals in Arctic waters. One of their favorite foods is shrimp, but they also feed on small fishes and zooplankton, which are tiny drifting animals.

eaten by

Shrimp

Shrimp and fishes feed on even smaller creatures called copepods.

eaten by

Copepods

These tiny cousins of lobsters and crabs are food for many larger animals, including shrimp and humpback whales. Copepods eat floating plants called phytoplankton.

eaten by

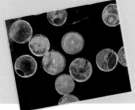

Phytoplankton

Microscopic plants, or phytoplankton, are sometimes called "the grass of the sea."

Tracking Changes

Studying indicator species is especially important now that scientists are watching the Arctic closely to see if it's getting warmer. What difference does one or two degrees of temperature change make? One way to find out is to look at an indicator species for clues.

Polar bears are an indicator species. In 1992 polar bears had a very good year. Scientists at the Churchill Northern Science Center in Manitoba, Canada, noticed that more cubs than usual were born that year and more survived. The Arctic weather was also slightly colder that year. Was there a connection?

Researchers think so. Polar bear cubs are born during the winter and spend the first months of their lives on the ice, where there are a lot of seals, polar bears' main food. Once the ice breaks up in the spring, the bears must move to land, where there are no seals, and they often go hungry.

Where do polar bears fit in a food chain? Right at the top. Studying polar bears can tell researchers about other animals that are lower in a food chain.

Colder temperatures made ice form earlier and break up later. That gave baby bears more time to get fat and strong before they had to move to land. What caused the change? The answer is in the Pacific Ocean. Thousands of miles away, in the Philippine Islands, a volcano erupted earlier that year. Soot in the air was swept northeastward until it reached the Arctic. There, the soot created a cloud that delayed the breaking up of the sea ice.

The cloud from the volcano was a onetime event. Scientists wonder what might happen if there were a permanent temperature change, as some researchers think could happen. If the Arctic warmed up for good, for example, what would happen to the tiny copepods? "At first they might get fatter," Carin says. "There would be more algae, and more food for them."

But warmer water and air might mean that new species could live in the Arctic, and they could take food away from the Arctic copepods. A longer summer might also mean too long a hungry period for polar bears.

In the Bering Sea, Cynthia, a whale biologist, is already studying changes in the movements of whales. Large numbers of whales have been seen in regions where they were not observed decades ago. Because whales need to go where there is a lot of zooplankton or fishes for them to eat, the change in the migration, or movement of the whales from one region to another, may reflect large

Warmer Arctic temperatures might mean different migration routes for seals and walruses that stay near the ice's edge.

changes in the ecosystem. "That doesn't mean that the change is permanent," Cynthia says. She's watching and waiting to see what the ice, the zooplankton, and the whales will do next.

In the Arctic, where the climate is under close watch for change, animals may give the first clues of what that change might mean. Watching indicator species, whether they are tiny copepods, towering polar bears, or giant whales, can give scientists important clues about what's changing in the Arctic and what those changes might mean for the animals that live there.

Some whales, such as beluga whales and bowhead whales, move with the ice's edge because that's where the food is. If more ice melts, what will that mean for these whales?

5 Hidden Mysteries

The Arctic is important to the world because it helps mix the planet's waters. Warm saltwater comes into the Arctic from oceans all over the world and cold freshwater flows back out in a never-ending exchange. Researchers, such as Keith Von Der Heydt (above), study the Arctic and its connection to the rest of the world.

So the Arctic seems to be getting warmer, and the warming may affect the creatures that live there. To help understand these changes, scientists are studying both the animals that live in the Arctic and the Arctic ice itself.

Why is the Arctic becoming warmer? Will Arctic warming affect the rest of the world? These are big questions. To answer them, scientists need to understand more about how the Arctic works.

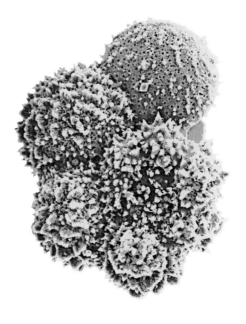

(left)
Live foraminifera look like this.

(below)
When you first see foraminifera shells, they look like sand. But when you see them under a microscope, like these fossil foraminifera that have been magnified 300 times, their appearance is amazing.

Sometimes answers to questions can be found in unexpected places, even in hidden places. Al Plueddemann and Dorinda "Rindy" Ostermann are two WHOI scientists who like to study things that are very hard to see.

Al studies waves under the Arctic ice, while Rindy studies tiny, one-celled creatures called foraminifera (pronounced "fo-ram-uh-NIF-ahr-uh"). Foraminifera shells collected from the bottom of the ocean give her clues to the temperature of the Arctic Ocean hundreds and thousands of years ago.

The hidden things that Al and Rindy study are signs of the way the Arctic Ocean works. Their research requires special equipment, careful planning, and the ability to imagine things that can't be seen easily. Why is it so important to understand the Arctic? Because the Arctic affects the rest of the world's oceans, and that affects everyone.

Arctic Investigation

Al Plueddemann investigates the invisible using special instruments that measure water motion. The instruments are hung beneath buoys like the one shown at right. Researcher Susumu Honjo developed these buoys to survive in the Arctic pack ice for several years. As each buoy drifts with the ice, it records temperature, salt levels, and ice thickness. Al's instruments gather information about the movement of water and waves under the ice.

Below, researcher Rick Krishfield (right) and two crew-members of the icebreaker *Polar Stern* lower Al's instruments into the water through a hole in the ice. The instruments will then be attached to a buoy. The buoy "plugs" the hole, eventually freezing into the ice.

Waves Beneath the Ice

The Arctic Ocean is like a giant mixing bowl. Warm saltwater comes into the Arctic, and cold freshwater flows out. Cold, warm, fresh, salty—what does it matter? It makes a difference. This exchange of cold and warm water and fresh and saltwater helps keep the world's oceans in balance so they are not too cold, too warm, or too salty.

Al studies a special kind of wave that helps mix Arctic waters. When most people think of waves, they think of the ones that travel along the ocean surface and break on the beach. But there is another kind of wave in the ocean that travels and breaks beneath the surface. In the Arctic these waves are found beneath the ice.

Imagine you hold a stick in a dish of water and jiggle it back and forth. Waves radiate, or spread out, from the stick. Now imagine holding the stick still and moving the dish of water back and forth instead. Waves will still radiate from the stick as the moving water flows around it. This is similar to what happens under the ice. Tidal currents move water back and forth, like the water in the dish. Bumps and ridges on the ocean floor act like the stick. As the tidal currents flow around the bumps and ridges, waves are radiated. These waves start at the bottom of the ocean and can move upward and side to side.

Under certain conditions, these underwater waves can break, similar to waves breaking on the beach. But underwater waves break beneath the ocean's surface. When they do, they help mix layers of cold freshwater with layers of warmer saltwater.

Underwater waves may be generated at the bottom anywhere in the ocean, but there are no top-generated underwater waves in the Arctic. Top-generated waves can be caused by a powerful storm disturbing water at the top of the ocean. But ice protects the top of the water in the Arctic. Because there are no top-generated waves in the Arctic, mixing by bottom-generated waves is particularly important. Studies like Al's can help scientists learn more about the Arctic and more about how the world's oceans work.

Underwater waves form when tidal currents move back and forth past bumps or ridges at the bottom of the ocean. The waves radiate upward from the bump and may eventually break beneath the ocean's surface, causing the mixing of cold freshwater and warmer saltwater.

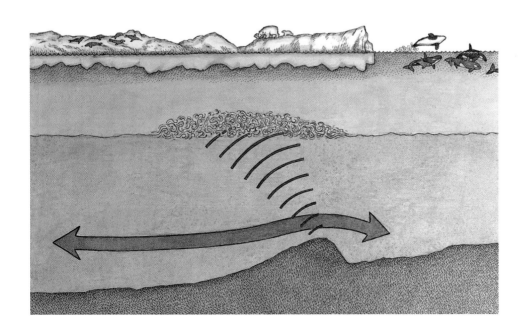

Scientists on research vessel *Bjarni Saemundsson* travel over the open Arctic Ocean to drop devices like this giant trap into the sea. The trap will remain anchored to the ocean bottom for up to a year. It will provide scientists like Rindy Ostermann with valuable samples, such as the shells of ocean animals, for study.

The shells in these bottles may help Rindy know if studying foraminifera shells is a good way to learn what the ocean's temperature was in the past.

What's the Temperature?

While Al is interested in knowing how the Arctic Ocean affects the rest of the world, Rindy Ostermann wants to know more about why the Arctic is warming up. There have been times in the past when the Arctic was warmer. Rindy collects evidence of these times by studying the shells of foraminifera living today and foraminifera fossils. Using the shells, Rindy may be able to tell what the temperature of the Arctic Ocean was hundreds, even millions, of years ago.

But first Rindy needs to determine if foraminifera shells are actually good thermometers. If new shells show correct current water temperatures, Rindy can use old fossil shells from the seafloor to calculate the Arctic water temperatures of long ago.

Rindy can tell temperatures from the shells because there are different types of foraminifera, and each type does best in water of a certain temperature. So a place

Sailing in the Arctic Ocean can be a tough job all by itself. On research vessel Knorr, scientists and the ship's captain and crew all work to break ice off the deck with mallets. (Just another day of science research in the Arctic!)

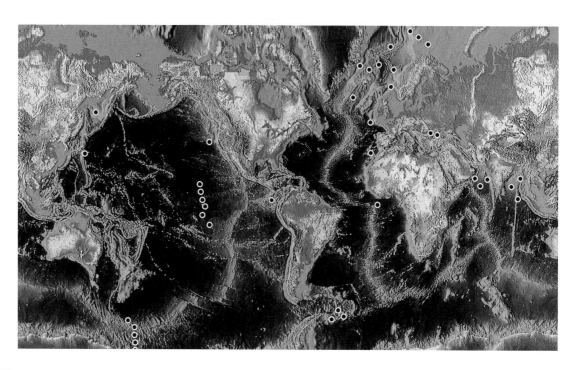

To help them organize their information, scientists often create a map or graph of their data. A computer helped Rindy make this map showing where traps have collected foraminifera.

where Rindy finds mostly warm-water foraminifera fossils, for example, is a place where the water was warm when those animals were alive.

Research starts on the open sea. Every May for the past 13 years, Rindy has sailed north from Iceland on a research vessel, setting down traps to collect the shells of foraminifera. Rindy's traps are funnels as tall as she is. Each trap collects shells from foraminifera that lived in the ocean during the year. As they drift down the funnel, the shells empty into bottles, with one bottle for each month. After Rindy collects a year's samples, each trap is returned to the seafloor to begin collecting again.

Back in her lab, Rindy studies the different kinds of shells in each bottle. By determining which kinds of foraminifera were collected in a particular place and making measurements on a special machine, Rindy can learn the water temperature when the animals were alive. So far the tiny shells have proved to be accurate water thermometers.

Rindy also collects fossil shells in cores taken from deep beneath the seafloor. She then determines the kind of foraminifera and the water temperature at the time the animals were alive. This kind of information about ancient Arctic conditions can help scientists determine if Arctic warming is part of a long, natural cycle, or if it has another cause, such as global warming caused by greenhouse gases.

The Arctic is home for a thriving ecosystem of plants, animals, ice, and water. It is also important as a mixing bowl for the world's oceans and as an indicator of the health of the planet.

Arctic scientists dream of going north. In the winter working in the Arctic means never seeing the sun. In the summer it means never seeing the sun go down. Sometimes it means homesickness, polar bear visits, or frostbite. At any time it means cold weather, deep silence, and an enormous sky. Every time it means new questions, new answers, new discoveries. The Arctic is a place scientists will be investigating for a long time to come.

Science in the Arctic can be difficult, cold, and even lonely, but it's also an unforgettable experience for those ready to take on the challenges of the far, far north.

Glossary

atmosphere [AT-mus-feer] The air surrounding Earth.

biologist [buy-OL-oh-jist] Someone who studies biology, the science of life.

buoy A device that floats on the water's surface or is frozen into ice.

copepod [CO-peh-pod] A tiny member of the crustacean family, related to shrimp and lobsters.

current [CUR-rent] A strong flow of water within the ocean.

ecosystem [EEK-oh-sis-tem] How living and nonliving things and their environment function together.

engineer [en-jin-EER] Someone who designs, builds, and tests machines and instruments.

food chain A group of living things arranged in order of predation, or what eats what. Phytoplankton are at the bottom of one Arctic food chain, and polar bears are at the top.

foraminifera [fo-ram-uh-NIF-ahr-uh] Tiny, one-celled marine animals with shells. Their shells provide clues to changing conditions in the Arctic.

fossil [FOSS-ul] Remains, such as a print or skeleton, of ancient plants or animals.

glacier A large mass of ice formed from compressed snow that moves slowly over land.

global warming A theory that pollution is increasing levels of greenhouse gases in the atmosphere. This could cause a rise in temperatures on Earth.

greenhouse gases Gases, mostly carbon dioxide but also including chemicals released by air conditioners, refrigerators, and some spray cans, that build up in the atmosphere and trap sunlight.

icebreaker A ship specially built and equipped to travel through ice, breaking through the ice pack as it goes.

ice cap Thick, permanent masses of ice that accumulate over land near the North and South Poles.

ice core A cylinder-shaped sample of ice, mud, or soil.

ice floe A large piece of sea ice that breaks and floats away from the ice pack.

indicator species [IN-di-kate-ur SPEE-shees] An animal or plant that is typical of an ecosystem. Looking at what is happening to an indicator species can provide clues about what is happening to other plants and animals in that area.

krill [KRIL] A member of the crustacean family (which includes lobster, crabs, and shrimp). Krill is a main food for seals and whales.

magnify [MAG-ni-fy] To make something appear larger by looking at it through a special lens, camera, or microscope.

microscopic [my-krow-SCOP-ick] So small as to be difficult or impossible to see without the help of a microscope or another magnifier.

oceanographer A scientist who studies the ocean.

pack ice Ice that freezes from seawater at the Poles to form masses with irregular borders that can either stick together or break apart.

underwater waves Waves that form and move under the ocean's surface.

water vapor [VAY-pur] The gas form of water.

Further Reading

Bramwell, Martin, and Marjorie Crosby-Fairall. *Polar Explorations: Journeys to the Arctic and Antarctic.* New York: Dorling Kindersley, 1998.

Curlee, Lynn. *Into the Ice: The Story of Arctic Exploration.* Boston: Houghton Mifflin, 1998.

Dabcovich, Lydia. *The Polar Bear Son: An Inuit Tale.* New York: Clarion, 1997.

Heinz, Brian J. Illustrated by Jon Van Zyle. *Kayuktuk: An Arctic Quest.* San Francisco: Chronicle, 1996.

Kalman, Bobbie. *Arctic World* Series. New York: Crabtree, 1998.

Markle, Sandra. *Super Cool Science: South Pole Stations, Past, Present, and Future.* New York: Walker & Company, 1998.

Rau, Dana Meachen. *Arctic Adventure: A Story of Inuit Life in the 1800s.* Washington, D.C.: Smithsonian Odyssey, 1997.

Silver, Donald M. *Arctic Tundra: One Small Square.* New York: McGraw Hill, 1997.

Steger, Will, and Jon Bowermaster. Illustrated by Allison Russo. *Over the Top of the World: Explorer Will Steger's Trek Across the Arctic.* New York: Scholastic, 1997.

Young, Karen Romano. Illustrated by Brian Shaw. *The Ice's Edge: The Story of a Harp Seal Pup.* Washington, D.C.: Soundprints/Smithsonian, 1996.

Index

Acknowledgments

The author would like to thank Carin Ashjian, James Bellingham, the Burnham family, Rick Krishfield, Jeff Lord, Kate Madin, Dorinda Ostermann, W. Brechner Owens, Albert J. Plueddemann, Cynthia Tynan, Keith Von Der Heydt, Dana Yoerger, and especially Samuel Young and his inspiring polar bear, Nansen.

Photographs courtesy of:

Alexander, B. and C./Photo Researchers, Inc.:32 second from top; Allan, Doug/British Broadcasting Corporation (BBC): 35; Ashjian, Carin/WHOI: 29 bottom, 32 second from bottom; Canadian Museum of Civilization: 14 left; Corbis/Bettmann: 3 bottom, 16 second from left, 16 top right; Corbis/Hulton-Deutsch Collection: 15 left, 16 left, 16 lower right; Corbis/Leonard de Selva: 12; Corbis/Library of Congress: 15 right, 16 top left; Corbis/Peter Harholdt: 14 middle; Corbis/Underwood and Underwood: 16 third from left; Foott, Jeff/BBC: 32 top; Freund, Jergen/BBC: 34; Gibbs, Wayt/Scientific American: 22, 24; Hines, Sandra/University of Washington: 1, 17 bottom right, 25, 28, 29 top; ITAR-TASS: 17 top right; Kadan, Susan: 16 lower left; Krishfield, Rick: 38; Lord, Jeff/WHOI: 3 top, 6, 10 bottom; Madin, Larry: 32 middle and bottom; Mangelsen, Thomas D./BBC: 33; Mark Twickler/University of New Hampshire: 26–27; Norris, Richard/WHOI: 37 middle and bottom; National Library of Norway: 13, 18; Ostermann, Dorinda/WHOI: 17 bottom left, 37 top, 40, 41 top, 42; Rowell, Galen: 4, 8 top; Submarine Force Museum, Groton, CT: 17 top left; Tupper, George/WHOI: 41; Tynan, Cynthia/National Oceanic Atmospheric Administration/National Marine Mammals Laboratories: back cover, 30, 31; University Museum of Cultural Heritage, University of Oslo: 14 right; University of Washington: 2, 20; Von Der Heydt, Keith/WHOI: 8 bottom, 9, 10 top, 11, 36, 42; Zimmerman, Sarah: front cover.

The illustration on pages 5, 15, and 19 is by David Stevenson.

Illustrations on pages 7, 21, and 23 are by Patricia Wynne.